FOLKLORE GUIDE TO THE WEATHER

by David Bowen, F.R.Met.S.

INDEX

	PAGE
Please read this first	2
TELL THE WEATHER from CLOUDS	3
,, DEW	5
,, MIST	6
,, HAZE	6
,, FOG	7
,, WIND	7
,, STARS, MOON, LUNAR HALO	9
,, SUN, MOCK SUNS, SOLAR HALO	10
,, RAINBOWS	11
,, RAIN	11
,, HAIL	13
,, THUNDER AND LIGHTNING	13
,, FROST	14
,, SNOW	15
,, SOUND	15
,, TEMPERATURE	16
,, BIRDS	16
,, ANIMALS (QUADRUPEDS)	19
,, INSECTS	22
,, FISH, MOLLUSCS AND REPTILES	23
,, PLANTS AND TREES	26
,, VARIOUS SIGNS	27
,, THE BAROMETER	30
For further reading	32

wyrd

Published by Wyrd Books,
an imprint of Read & Co.

This edition published by Read & Co. in 2017

Extra material © 2017 Read & Co. Books

All rights reserved. No portion of this book may be reproduced in any form without the permission of the publisher in writing.

A catalogue record for this book is available from the British Library.

ISBN: 9781528700191

Read & Co. is part of Read Books Ltd.
For more information visit www.wyrdbooks.co.uk

Please read this first . . .

MOST BOOKS ABOUT THE WEATHER devote a few pages to some of the weather sayings of folklore, and there are one or two books that give more than just a handful of them. But in almost every case the good ones seem to be mixed with the bad.

It is the aim of this **GUIDE**, however, to give only those sayings, which, in the author's opinion, are reliable at least seven times out of ten. They are arranged in convenient groups and can be referred to at a moment's notice.

Weather sayings today are generally distinguished by whether or not they can be backed up by scientific explanations. But we should remember that the mere absence of an explanation does not necessarily mean that a particular saying is unreliable.

Some of the most reliable sayings, according to countrymen, are those which interpret the behaviour of birds and insects, yet only a few of these can be proved scientifically. We can only assume that small creatures—obviously more dependent on the weather than human beings—are extra sensitive to weather changes, and even to the minute changes in the atmospheric conditions that precede what we regard as a change in the weather. To us, therefore, they may be prophets, even if for their own part they are only responding to the existing conditions.

Most of the really unreliable sayings can be easily recognised, for these are the ones that are obviously based on little more than superstition. But even these serve a useful purpose if it can be argued that, by encountering them, we are sharpening our weather wits.

How is it, people ask, that the unreliable sayings still linger on?

Mainly, I think, because even the worst of them have a fifty-fifty chance of working out.

There are literally tens of thousands of weather sayings altogether, many of them a good deal older than Christianity. Some of the very old ones can still be used today, both in and away from the countries in which they originated. Others, including many British ones, are only valid within a small local area. Those quoted in this **GUIDE** are very straightforward and can be used almost anywhere, and the few that have a conditional or a rather restricted use are marked accordingly.

The **GUIDE** has been designed for the pocket rather than the bookshelf, and, it is hoped, will be taken for many a walk into the country. It should help the reader, no matter where he happens to be, to interpret his local weather for himself.

Clouds . . .

Small and white, like fleeces
>If woolly fleeces spread the heavenly way,
>Be sure no rain disturbs the summer day.

Large, cliff-like
>When mountains and cliffs in the clouds appear,
>Some sudden and violent showers are near.

According to colour and outline
>Light, delicate, quiet tints or colours, with soft, undefined forms of clouds, indicate and accompany fine weather ; but unusual or gaudy hues, with hard, definitely outlined clouds, foretell rain, and probably strong wind.

Capping hills
>When mountains and hills appear capped by clouds that hang about and embrace them, storms are imminent.

Bank in West
>A bench (or bank) of clouds in the west means rain.

Stationary, Piling up
>When clouds are stationary and others accumulate by them, but the first remain still ; it is a sign of a storm.

Settling back
>When a heavy cloud comes up in the south-west, and seems to settle back again, look out for a storm.

Soon collecting
>If the sky, from being clear, becomes quickly fretted or spotted all over with bunches of clouds, rain will soon fall.

" Mackerel "
>A mackerel sky denotes fair weather for that day, but rain a day or two after.

Curdled
>A curdly sky
>Will not leave the earth long dry.

CLOUDS (continued from previous page) .

" *Painter's brush*," " *Goat's hair*," " *Mare's tail* "
>Trace in the sky the painter's brush,
>Then winds around you soon will rush.

(*This is the high cloud called ' goat's hair ' or the ' gray mare's tail,' foreboding wind and rain. It appears in tufts bunched closely together, and is not to be confused with the " threadlike " high clouds of fine weather.*)

" *Hens' scratchings* "
>Hens' scarts (scratchings) and filly tails
>Make lofty ships carry low sails.
>(*Same cloud as the previous one*)

High sheet, gloomy

A high sheet of cloud spreading across the whole sky, and casting a general gloom over the countryside, presages rain and wind.

Two Layers

If two layers of cloud appear in hot weather to move in different directions, they indicate thunder.

If, during dry weather, two layers of cloud appear moving in opposite directions, rain will follow.

Dark and heavy

Dark, heavy clouds, carried rapidly along near the earth, are a sign of great disturbance in the atmosphere from conflicting currents. At such times the weather is never settled, and rain extremely probable.

From south, during frost

If, during a frost, clouds drive up high from the south, expect a thaw.

Early disappearance

If at sunrising the clouds are driven away, this denotes fair weather.

When overhead or otherwise

When it is bright all round it will not rain ; when it is bright only overhead it will.

Sun tinted (1)
> The evening red and the morning grey
> Are the tokens of a bonny day.

(The saying refers to a hazy-red evening and fleecy early morning mist.)

Sun tinted (2)
> A red morn . . . ever yet betokened
> Wreck to the seaman, tempest to the field
> Sorrow to shepherds, woe unto the birds,
> Gust and foul flaws to herdmen and to herds.

Zinc-grey layer
A low zinc-grey layer covering the whole sky is a sign of snow. If the temperature is around 37 deg. Fahrenheit, the flakes will be large ; if much below 37 deg. they will be small.

Green
If the clouds take on a greenish hue, a heavy storm of rain is very imminent.

Milky
> If the sky looks washed with a milky white,
> The rain is near though not yet in sight.

No clouds
A very clear sky without clouds is not to be trusted, unless the barometer be high.

Dew . . .

Before midnight
> With dew before midnight,
> The next day will sure be bright.

(This is confirmed if the dew is still plentiful in the early morning.)

No dew
If in clear summer nights there is no dew, expect rain next day.

Mist . . .

From sea
> When the mist comes from the sea,
> Then good weather it will be.

Rising up hill
> Thin, white, fleecy, broken mist, slowly ascending the sides of a hill or of a mountain whose top is uncovered, predicts a fair day.

Rising from low ground
> If mists rise in low ground and soon vanish, expect fair weather.

Spreading over rivers
> A white mist in the evening, over a meadow with a river, will be drawn up by the sun next morning, and the day will be bright.
> (*The rivers or marshes may actually be seen to steam.*)

Falling on hillsides
> Thin white mist, which settles on the hillsides after a warm day in summer, is a sign of continued fair weather.

With wind
> See *Fog, with wind*.

Haze . . .

In winter
> A persistently hazy atmosphere in winter is a sign of cold raw weather.

In summer
> The greater the haze, the more settled the weather.

Fog . . .

Falling, in autumn and winter
   Predict fog in autumn and winter when (1) the sky is clear (or clears) at sunset ; (2) there is no more than a mere breath of wind ; (3) when the air is fairly damp.

Successive foggy nights
   Three foggy nights in a week ; then expect foggy days as well.

Hanging below trees
   Heavy fog in winter, when it hangs below trees, is followed by rain.

Damp with wind
   If there be a damp fog or mist, accompanied by wind, expect rain.

Becoming damper
   An originally dry fog that is becoming gradually damper indicates rain, and probably wind.

With slight breeze
   During a thick town-fog a breath of air on the face, followed by a slight swirling, is generally the first sign of a clearance.

In summer
   A summer fog is for fair weather.

Wind . . .

Light breezes
   The smaller and lighter winds generally rise in the morning and fall at sunset.

Sudden gusts
   Sudden gusts never come in a clear sky, but only when it is cloudy and with rain.

Sudden freshening
   The sudden storm lasts not three hours.

WIND (continued from previous page) . . .

Veering and backing
>A veering wind, fair weather.
>A backing wind, foul weather.

(*The wind " veers " when its change of direction is clockwise, e.g., when it blows from the west after blowing from the south. It " backs " when the change is the other way about.*)

With rain
>When rain comes before wind,
>Halyards, sheets, and braces mind ;
>But when wind comes before rain,
>Soon you may make sail again.

With frost
>When the hoar-frost is first accompanied by east wind, it indicates that the cold will continue a long time.

North-west
>A north-westerly gale
>Brings showers of hail.

North-west compared with north-east
>North-west wind brings a short storm ; a north-east wind brings a long storm.

Northerly
>Northerly winds bring showers rather than continuous rains or snow.
>>The north wind doth blow
>>And we shall have snow.*

(**Snow flurries or showers. Continuous snow does not generally fall unless the wind is between east and south-east in winter and the temperature approximately* 37 *deg. Fahrenheit.*)

West
>A western wind carrieth water in his hand.

South, in summer
>The south wind, when gentle, is not a great collector of clouds during the summer months, but if it becomes violent, it makes the sky become cloudy and brings on thunder and rain.

South, in winter
　The south wind, during the winter months, will bring mild, cloudy weather, with drizzle.

South, when showery
　　　A southerly wind with showers of rain
　　　Will bring the wind from west again.

Unsteady
　Unsteadiness of wind shows changing weather,

Frequent change, with agitated clouds
　A frequent change of wind, with agitation in the clouds, denotes a storm.

Making a swell
　　When there's hardly a wind,—but a swell on the sea,
　　For certain a drench and a gale there will be.

Stars, Moon, Lunar Halo . . .

Many stars
　When the sky seems very full of stars expect rain the following day, or, in winter, frost.

Moon sharp, with horns
　When the moon's horns are sharp and well defined, expect rain the following day, or, in winter, frost.

Reddish moon, through haze
　If the moon appears a reddish brown through the haze, the weather will stay fair.

" Average " moon
　If the moon is distinct, neither too sharp in outline nor, on the other hand, " watery " and blurred, the weather will stay fair for the time being.

Pale moon
　If the full moon rises pale, expect rain.

Lunar halo
　　　If the moon rises haloed round,
　　　Soon you'll tread on deluged ground.

Sun, Mock Suns, Solar Halo . . .

Bright morning sun
>The night has gone by,
>'Tis a bright sunlit sky,
>But there's rain by and by
>If the glass is not high.

Sun reflected on clouds
>A red sky in the morning
>Is the shepherd's warning ;
>Though a red haze at night
>Is the shepherd's delight.

(*The red sky referred to is angry red in colour* ; *the same sky in the evening is also a sign of rain. It is the lighter hazy-red sky which indicates fair weather.*)

Sun overhead, and on all sides
>Sun blazing on high
>And clouds draped all round—
>You won't stay long dry
>'Less shelter you've found !
>Yet with clouds overhead,
>If there's sunshine all round
>The rain that is shed
>Will not soak the ground.

Pale sun
>If the sun's shining pale with a watery eye,
>Be sure of a soaking ere nightfall is nigh.

Mock suns
>Mock suns presage a certain change in the weather, and generally rain or heavy showers the same day.

Solar halo
>When the sun is in his house (halo), it will rain soon.

Rainbows . . .

Morning and night

> A rainbow in the morning
> Is the shepherd's warning;
> A rainbow at night
> Is the shepherd's delight*.

(*The "delight," in this case, is not always long-lived. The rainbow in the evening denotes a clearing sky, but not necessarily a change to the better for the day following the evening.)

Windward and leeward

> Rainbow to windward, foul falls the day;
> Rainbow to leeward, damp runs away.

(*This saying can be used in conjunction with the one above. It is very seldom that the two will appear to contradict each other*).

Suddenly disappearing

If, when the clouds are not thick, the rainbow forms and disappears suddenly, the prismatic colours being but slightly discernible, expect fair weather next day.

Rain . . .

During the morning

> Rain before seven,
> Clear by eleven.

(*Because most rainbelts do not give more than 4-5 hours of rain to any area. The clearance, however, may be only short-lived. This saying can be adapted for any part of the day: e.g., "Rain before twelve, clear by four."*)

RAIN (continued from previous page) . . .

From the east
>When the rain is from the east,
>It lasts a day or two at least.

(*Fortunately most rain-belts approach from a point between south-east and north-west.*)

From the west
>When rain comes from the west, it will not be more than a few hours before the weather improves, and becomes brighter, but showers are likely to follow.

From the north-west
>With the rain of the north-west, expect showers of hail.

Becoming sharper
>A sharp shower of rain following a period of light—but continuous—rain or drizzle is a sign that the weather will soon improve.

Followed by hail
>If hail appears after a long course of rain, it is a sign that the weather will improve very soon indeed. (See Hail).

Sudden, and long warning
>Sudden rains never last long ; but when the air grows thick by degrees, and the sun, moon, and stars shine dimmer and dimmer, then it is likely to rain for four or five hours, and sometimes even longer.

Showers and sunshine
>Showers and sunshine, the two in their turn,
>Bring certain good weather for which we do yearn.
>But showers and sunshine, then gloom overhead,
>Will bring on more rain for some hours, it is said.

Light showers during a drought
>If very light, short showers come during dry weather, they are said to ' harden the drought ' and indicate no change.

In winter, followed by a sudden frost
>Sudden frosts, in winter, after rain, soon bring back more rain again.

Hail . . .

After rain
> Hail after a long course of rain is a sign that the weather will improve very soon indeed, but showers will follow if the wind is westerly, and the stronger the wind the sharper the showers—which may be of hail.

By day
> A hailstorm by day denotes a frost at night.

Thunder and Lightning . . .

Distant thunder
> The distant thunder speaks of coming rain.

Much thunder
> After *much* thunder, *much* rain.

Thunder from south and north
> Thunder from the south or south-east indicates long storms; from the north or north-west, short storms.

Thunder morning and noon
> When it thunders in the morning, it will rain before night. Thunder in the morning denotes winds; at noon, showers.

Thunder and lightning in summer
> Thunder and lightning in the summer show
> The point from which the freshening breeze will blow.

Lightning from one or more directions
> Lightning signifies the approach of wind and rain from the quarter where it lightens; but if it lightens in different parts of the sky, there will be severe and dreadful storms.

With large or small fall of barometer
> A thunderstorm that is accompanied by a large fall of the barometer will be very severe; but it will be slight and merely local in extent if there is little or no fall of the barometer.

Frost . . .

Following rain

Frost suddenly following heavy rain seldom lasts long.

Before and after snow

In winter, during a frost, if it begins to snow heavily, the temperature of the air generally rises to 37 deg. (or near it), and continues there while the snow falls ; after which, if the weather clears up, expect severe cold.

Heavy

Heavy frosts are generally followed by fine, clear weather.

Breaking

Signs of frost breaking up :
1. The sun looking watery at rising.
2. The sun setting in bluish clouds, and casting reflected rays into them.
3. The appearance of high clouds driving up from the south.
4. The stars looking dull, and the moon's horn's blunted.
5. A fall of the barometer.

Becoming more intense

Expect the frost to increase in severity, and the weather to become drier and crisper, if, in winter, the wind veers from north-west to north-east.

(*The wind " veers " when its change of direction is clockwise, and " backs " when the change is the other way about.*)

With rise of the barometer

A rise of the barometer during a frost will increase the cold.

Snow . . .

Falling or melting, according to the temperature
>Dry all the flurries of granular snow,
>When the mercury's zero and falling below ;
>Sticky the snowflakes that pile up in drifts,
>If now up to thirty-and-seven it shifts ;
>And oh for a thawful of dirt and of slush,
>If *past* thirty-seven it soars in its rush !

Presaging colder or warmer weather, according to size
If the snowflakes decrease in size and become drier, the weather will become colder ;. if they increase in size, it will become slightly warmer ; if they increase very considerably in size there will soon be a thaw.

Passing or drifting, according to wind direction
>Snow from the north and north-west
>Is the driest and will not molest ;
>But coming from the east and south-east,
>'Twill drift o'er both man and his beast.

Sound . . .

Good day for hearing
A good hearing day is a sign of wet.

Noise from forest and mountain
>If forests all murmur and mountains do roar,
>Then close all your windows and bolt up your door.

Echo on the shore
The shores sounding in a calm, and the sea beating with a murmur or an echo louder and clearer than usual, are signs of wind and rain.

Bells
The ringing of bells is heard at a greater distance before rain ; But before wind it is heard more unequally, the sound coming and going.

Temperature . . .

Falling steeply
> If the temperature falls steeply in summer, expect unsettled weather ; if it falls steeply in winter expect hard, often sunny weather, or unsettled rainy weather according to whether the barometer is high or low respectively.

Falling gradually, in winter
> In autumn and winter, when the barometer is high, a gradual fall of temperature will bring on a fog.

Damp and dry heat
> When there is very close, hot weather, there is generally a reaction, and a severe storm follows, but the weather will stay fine so long as the heat is not damp.

Rising in summer
> A rising thermometer in the summer is a sign of settled weather, particularly if the barometer is high and steady.

Rising in winter
> If the temperature rises during the winter, it is a sign of rain ; it is generally accompanied by a falling barometer.

Birds . . .

Swallows
> Swallows high,
> Staying dry ;
> Swallows low,
> Wet 'twill blow.

Larks
> If larks fly high and sing long, expect fine weather.

Rooks
> If rooks feed in the streets of a village, it shows that a storm is near at hand.
>
> When rooks seem to drop in their flight, as if pierced by a shot, it is said to foreshow rain.
>
> Rooks will not leave their nests in the morning before a storm.

Ravens
>Ravens, when they croak continuously, denote wind ; but if the croaking is interrupted or stifled, or at longer intervals, they show rain.
>
>When ravens sit in the sun, expect fine weather to last.
>
>If the raven makes several different cries in the winter, it is a sign of storm.

Crows
>The continual prating of the crow, chiefly twice or thrice quick calling, indicates rain and stormy weather.

Blackbirds
>When the voices of blackbirds are unusually shrill, rain will follow.

Robins
>If a robin sings on a high branch of a tree, it is a sign of fine weather ; but if one sings near the ground, the weather will be wet.

Thrushes
>When the thrush sings at sunset, a fair day will follow.

Woodpeckers
>When woodpeckers are much heard, rain will follow.

Sparrows
>If sparrows chirp a great deal, wet weather will ensue.

Fowls
>If the fowls huddle together outside the henhouse instead of going to roost, there will be wet weather.
>
>If fowls grub in the dust and clap their wings, or if their wings droop, or if they crowd into a house, it indicates rain.

Cocks
>>If the cock goes crowing to bed,
>>He'll certainly rise with a watery head.
>
>If cocks crow during a downpour it will be fine before night.

BIRDS (continued from previous page) . . .

Guinea-fowls
Guinea-fowls squall more than usual before rain.

Land Birds
Land birds are observed to bathe before rain.

Tree Birds; Herons
If birds that dwell in trees return eagerly to their nests, and leave their feeding-ground early, it is a sign of storms ; but when a heron stands melancholy on the sand it only denotes rain.

Herons
Herons in the evening flying up and down, as if doubtful where to rest, presage some evil approaching weather.

Geese
> The goose and the gander
> Begin to meander ;
> The matter is plain,
> They are dancing for rain.

Ducks, Geese
If ducks and geese fly backwards and forwards, and continually plunge in water and wash themselves incessantly,wet weather will ensue.

Geese
> If the wild geese gang out to sea,
> Good weather there will surely be.

Sea-gulls
> Sea-gull, sea-gull, sit on the *sand*,
> You bring bad weather flying o'er land.

Island birds
If in summer many birds which usually live on an island, appear in flocks (on the mainland), it indicates rain.

Peacocks
> When the peacock loudly bawls,
> Soon we'll have both rain and squalls.

Owls
Screech-owls are most noisy just before rain.

An owl hooting quietly in a storm indicates fair weather, and also when it hoots quietly by night in winter.

If owls hoot at night, expect fair weather.

Quails
When quails are heard in the evening, expect fair weather next day.

Petrels
The petrel is found to be a sure token of stormy weather. When these birds gather in numbers in the wake of a ship, the sailors feel sure of an impending tempest.

Turkeys
Turkeys perched on trees and refusing to descend indicate snow.

Ptarmigans
The frequently repeated cry of the ptarmigan low down on the mountains during frost and snow indicates more snow and continued cold.

Birds whistling, and silent
If birds begin to whistle in the early morning in winter, it is a sign of frost.

When birds are unusually silent in the day-time, it indicates thunder.

Animals [Quadrupeds] . . .

Animals sheltering, and crowding
When animals seek sheltered places instead of spreading over their usual range, an unfavourable change is probable.

If animals crowd together, rain will follow.

ANIMALS (QUADRUPEDS) (continued from previous page) . .

Cats
> When a cat sneezes, rain is near.
>
> When the cat scratches the table legs, a change of weather is coming.

Dogs
> Dogs making holes in the ground, howling when anyone goes out, eating grass in the morning, or refusing meat, are said to indicate coming rain.
>
> The unsusual howling of dogs portends a storm.

Horses and mules
> Horses and mules, if very lively without apparent cause, indicate cold.
>
> When horses lie with their heads upon the ground, it is a sign of rain.

Ponies
> Capering and scampering of wild ponies is a sign of rain, as it is also when they leave their moorland lairs and come down in droves to the low ground.

Pigs
> When pigs carry straw to their sties, bad weather may be expected.

Goats
> The goat will utter her peculiar cry before rain.
>
> Goats leave the high ground and seek shelter before a storm.

Asses
> If asses hang their ears downward and forward, and rub against walls, rain is approaching.
>
> If asses bray more frequently than usual, it foreshows rain.
>
>> Hark ! I hear the asses bray ;
>> We shall have some rain to-day.

Cattle
>When cattle remain on hill-tops, fine weather to come.
>
>When cattle lie down in the open during light rain, it will soon pass.
>
>When cows slap their sides (or a hedge) with their tails, it is a sign of rain.

Bulls, and bullocks
>If bulls lick their hooves or kick about, expect much rain.
>>The herdsmen too, while yet the skies are fair,
>>Warned by their bullocks, for the storm prepare—
>>When with rough tongue they lick their polished hoof,
>>When bellowing loud they seek the sheltering roof.

Oxen
>If oxen turn up their nostrils and sniff the air, or if they lick their forefeet, or lie on their right side, it will rain.

Sheep
>If sheep feed up-hill it is a sign of fine weather.
>>When sheep do huddle by tree and bush
>>Bad weather is coming with wind and slush.
>
>When sheep turn their backs to the wind, it is a sign of rain.
>
>If sheep gambol and fight, or retire to shelter, it presages a change in the weather.

Squirrels
>>Squirrels feeding on the tree,
>>Weather warm as warm can be.

Moles
>If moles throw up more earth than usual, rain is indicated.
>
>When the mole throws up fresh earth during a frost, it will thaw in less than forty-eight hours.

Bats
>When bats appear very early in the evening, expect fair weather; but when they utter plaintive cries, rain may be expected.
>
>If bats fly abroad longer than usual after sunset, fair weather.
>
>If bats abound and are vivacious, fine weather may be expected.

Insects . . .

Spiders

Before rain or wind spiders fix their frame-lines unusually short. If they make them very long, the weather will usually be fine for fourteen days.

When the spider cleans its web, fine weather is indicated.

If spiders break off and remove their webs, the weather will be wet.

If the spider works during rain, it is an indication that the weather will soon be clear.

Spiders, when they are seen crawling on the walls more than usually, indicate that rain will probably ensue.

If the spiders are totally indolent, rain generally soon follows.

Crickets

When crickets chirp unusually, wet is expected.

Small black insects

When little black insects appear on the snow, expect a thaw.

Ants

Expect stormy weather when ants travel in lines, and fair weather when they scatter.

.If ants their walls do frequent build,
Rain will from the clouds be spilled.

Ants withdraw into their nests and busy themselves with their eggs before a storm.

Bees

Bees will not swarm
Before a near storm.

Bees are restless before thundery weather.

When bees to distance wing their flight,
The days are warm and skies are bright ;
But when the flight ends near their home,
Then rain and cold are sure to come.

A bee was never caught in a shower.

Clock-beetles
    If the clock-beetle flies circularly and buzzes, it is a sign of fine weather.

Wasps
    Wasps in great numbers and busy indicate warm weather.

Flies
    If flies sting and are more troublesome than usual, a change approaches.

> When harvest flies hum,
> Warm weather to come.

If flies cling much to the ceilings, or disappear, rain may be expected.

> A fly on your nose, you slap, and it goes ;
> If it comes back again, it will bring a good rain.

Gnats
    If gnats fly in large numbers, the weather will be fine.

Woodlice
    If woodlice run about in great numbers, expect rain.

Fish, Molluscs and Reptiles . . .

Fish biting
    If, during damp rainy weather, fish bite readily and swim near the surface, an improvement is likely, or, if it remains cloudy, it will be quiet rather than windy.

(*On occasions fish have been known to bite before rain. But this is more usual after rain than before, as it is the rain that reoxygenates the water, energising the fish which have remained dormant. One thing is certain : fish will not bite when a major change in the weather is impending.*)

Pike
    When pike lie on the bed of a stream quietly, expect rain or wind, and in winter, cold weather.

FISH, MOLLUSCS and REPTILES (continued from previous page)

Trout
>When trout refuse bait or fly,
>There ever is a storm a-nigh.

Eels
>If eels become very agitated, it is a sign of rain.

Clam-beds
>Air bubbles over the clam-beds indicate rain.

Porpoises
>A school of porpoises in or near harbour is a sign of strong winds or storms approaching.

Sharks
>Sharks go out to sea at the approach of a wave of cold weather.

Sea-anemones
>The sea-anemone closes before rain, and opens for fine, clear weather.

Sea-urchins
>When sea-urchins thrust themselves into the sand, a storm is approaching.

Snails and slugs
>If snails and slugs come out abundantly, it is a sign of rain.

Snails
>Snailie, snailie, shoot out your horn,
>Point to a calm and bonny morn!

Toads and frogs
>When toads and green frogs creep out in large numbers, the weather will be wet.

Tree-frogs
>Tree-frogs piping during rain indicate its continuance. Tree-frogs crawl up to the branches of trees before a change of weather.

Frogs

When frogs croak much, it is a sign of rain.
The louder the frog, the more the rain.

Grass snakes

An abundance of grass-snakes is a sign of rain. They will also be seen nearer to houses before rain.

Worms

Worms descend to a great depth before either a long drought or a severe frost.

When common garden worms form many 'casts,' rain is likely ; many casts in winter after a night of frost is a sign of a thaw within twelve hours.

When many earth-worms appear at the surface, expect rain ; if they appear in the evening expect a mild damp night.

Glow-worms

If glow-worms shine more brightly than usual they indicate rain within forty-eight hours, more especially when they remain luminous a short time after mid-night, the hour at which they generally extinguish their lamps.

Leeches in a jar

The leeches remain at the bottom during absolutely fine (and calm wet) weather. When a change in the former is approaching, they move steadily upwards many hours, even twenty-four or rather more, in advance. If a storm is rapidly approaching, the leeches become very restless, rising quickly ; while previous to a thunderstorm they are invariably much disturbed, and remain out of the water. When the change occurs and is passing over, they are quiet, and descend again. If under these circumstances they rise and continue above water, length or violence of storm is indicated. If they rise during a continuance of east wind, strong winds rather than rain are to be looked for.

(*A leech in a jar was one of the chief meteorological attractions at the Great Exhibition of* 1851. *A cord, tied loosely to the leech, extended from the jar and rang a little bell above whenever the creature moved.*)

Plants and Trees . . .

Chickweed

Chickweed expands its leaves boldly and fully when fine weather is to follow ; but if it should shut up, then the traveller is to put on his great-coat.

The half opening of the flowers of the chickweed is a sign that the wet will not last long.

Scarlet pimpernel

The closing of the scarlet pimpernel's flowers in the day-time ' betokeneth rain and foul weather, contrariwise, if they be spread abroad, fair weather'.

Goat's beard

Goat's beard keeps its flowers closed in damp weather.

Dandelions

Dandelions close their blossoms before a storm.

When the down of the dandelion contracts, it is a sign of rain.

Colt's-foot, dandelion and thistles

If the down flies off colt's-foot, dandelion and thistles, when there is no wind, it is a sign of rain.

Whitlow grass and lady's bedstraw

We may look for wet weather if the leaves of the whitlow grass droop, and if lady's bedstraw becomes inflated and gives out a strong odour.

Trefoil

The stalk of trefoil swells before rain.

Clover

Clover contracts its leaves at the approach of a storm.

Sensitive plants

Sensitive plants contract their leaves at the approach of rain.

Wood anemone

The yellow wood anemone and the wind-flower (*Anemone nemorosa*) close their petals and droop before rain.

Pondweed

Pondweed sinks before rain.

Watercress beds
> If watercress beds steam on a summer evening the next day will be hot.

Pine-cones
> Pine-cones hung up in the house will close themselves against wet and cold weather, and open against hot and dry times.

Seaweed
> A piece of kelp or seaweed hung up will become damp previous to rain.

Undersides of leaves
> When the leaves show their under sides,
> Be very sure that rain betides.

Silver maple
> The silver maple shows the lining of its leaf before a storm.

May-tree
> The leaves of the may-tree bear up, so that the underside may be seen before a storm.

Lime, sycamore, plane, and poplar
> Before rain the leaves of the lime, sycamore, plane, and poplar trees show a great deal more of their under surfaces.

Aspen
> Trembling of aspen leaves in calm weather indicates an approaching storm.

Dead branches
> Dead branches falling in calm weather indicate rain.

Leaves and straws in the air
> Leaves and straws playing in the air when no breeze is felt, the down of plants flying about, and feathers floating and playing on the water, show that winds are at hand.

Various Signs . . .

Hills very distinct
> When the distant hills are more than usually distinct, rain approaches—if not the same day, then generally within twenty-four hours.

Soot falling and burning
> If a good deal of soot falls down the chimney, rain will ensue. Soot burning on the back of the chimney indicates storms.

VARIOUS SIGNS (continued from previous page) . . .

Soot hanging and sparkling
> Soot hanging from the bars of the grate, a sign of wind. When the soot sparkles on pots over the fire, rain follows.

Coals bright
> Coals, when they burn very bright, foretell wind, and likewise when they quickly cast off and deposit their ashes.

Candles
> When the flames of candles flare and snap or burn with an unsteady or dim light, rain and frequently wind also are found to follow. But when they burn with a soft and steady light the weather will be fine.

Smoke falling
> When you observe smoke from the chimney of a cottage descend upon the roof and pass along the eaves, expect rain within six hours.

Dust
> If dust whirls round in eddies when being blown about by the wind, it is a sign of rain.

Walls damp
> When walls are more than usually damp, rain is expected. When in cold weather the walls begin to show dampness, the weather changes.

Drains smelling
> Drains, ditches, and dunghills are more offensive before rain.

Springs rising
> Springs rise against rain.

Matting
> If the matting on the floor is shrinking, dry weather may be expected. When the matting expands, expect wet weather.

Floors damp
> Oiled floors become damp before rain.

Chairs and corns
> See how the chairs and tables crack,
> Old Betty's joints are on the rack ;
> Her corns with shooting pains torment her
> And to her bed untimely send her.
> 'Twill surely rain, I see with sorrow ;
> Our jaunt must be put off tomorrow.

Rheumatism
> When rheumatic people complain of more than ordinary pains in the joints, it will rain.

Depressed spirits; feet cold
> Persons of a nervous temperament have a sense of dread or a depression of spirits preceding a fall of rain. With some persons, the feet tend to go cold before snow, and the blood vessels relax when it falls.

Ears ringing
> When the ears ring, the weather will change.

Dreams
> Dreams of a hurrying and frightful nature are a sign of a change in the weather; and likewise imperfect and fitful sleep.

Appetite
> When everything at the table is eaten, it indicates continued clear weather.

Milk souring
> Cream and milk, when they turn sour in the night, often indicate thereby that thunderstorms are about.

Webs
> Spiders' webs scattered thickly over a field covered with dew glistening in the morning sun indicate continued fine weather.
> > When you see gossamer flying,
> > Be sure the air is drying.

Rag
> On the principle of the seaweed hygroscope, a piece of rag soaked in brine and allowed to dry will become damp in moist conditions of the air.

Catgut
> Strings of catgut or whipcord untwist and become longer during a dry state of the air, and *vice versa*.

Weather-house
> On this principle is constructed the weather-house—a toy from which the figure of a woman emerges in fine weather, and one of a man wrapped in a great-coat comes out before rain.

Plates
> If metal plates and dishes sweat, it is a sign of bad weather.

Camphor gum
> Camphor gum dissolved in alcohol will throw out feathery crystals before rain.

The Barometer . . .

Falling low; and high
> A glass falling low
> Will bring you a blow;
> But when steady and high
> You'll have bright cloudless sky.

(*Two warnings in connection with the above.* (1) *Between storms there is often a sudden rise of the barometer before it plunges again, so that only a steady rise foretells fine weather, and:*
> *Quick rise after low,*
> *E'en stronger the blow.*

(2) *Although a high barometer almost invariably brings calm weather, and bright and warm days in spring, summer and early autumn; in late autumn and winter there is the danger of fog.*)

Length of warning
> Long foretold, long last;
> Short notice, soon past.

(*i.e., a gradual fall of the barometer over a long period brings more hours of rain than a more sudden drop. The latter, however, is a sign of very squally weather, even if it does not last very long.*)

Low, after rain
> Should the barometer continue low when the sky becomes clear after heavy rain, expect more rain within a very short time.

Falling, winter and summer
> The barometer usually falls for southerly and westerly winds, and for damper, stormier, and warmer weather in winter, and for damper, stormier and cooler weather in summer.

Low, with clouds
> If there are large clouds in the sky when the barometer is low, or falling, expect rain to fall from them, and if the sky then beomes milky expect a long period of rain.

Falling, and wind backing
> When the wind backs and the weather glass falls,
> Then be on your guard against gales and squalls.

(*The wind " backs " when its change of direction is anticlockwise: e.g., when it blows from the south after blowing from the west.*)

Falling, for winds and rain
>   The barometer falls lower for high winds than for heavy rains. If the fall amount to one inch in twenty-four hours, expect a very severe gale.

Falling, wind southerly
>   A fall of the mercury with a south wind is invariably followed by very damp weather, and, in summer, by very heavy rain.

Falling, in summer
>   In summer, when the barometer falls suddenly, expect a thunderstorm ; and if it does not rise again when the storm ceases, there will be several days' unsettled weather.
>
>   A summer thunderstorm, which does not much depress the barometer, will be very local and of slight consequence.

Rising, ground dry
>   When the barometer rises considerably, and the ground becomes dry, although the sky remains overcast, expect fair weather within a few days, even if, for a time, it remains cool in those areas facing the prevailing wind.

Steady, after gales
>   When, after a succession of gales and great fluctuations of the barometer, a gale comes on from the south-west, which does not cause much, if any, depression of the instrument, you may consider that more settled weather is near at hand.

Falling, after a frost
>   The surest sign of a thaw after a long spell of frost is a steady fall of the barometer.
>
>   If, after a sudden frost, the barometer falls two or three tenths of an inch, expect a thaw.

Falling in winter, temperature rising
>   During winter heavy rain is indicated by a decrease of pressure and an increase of temperature. But heavy snow is indicated if the temperature increase is from the freezing point (32 degrees Fahrenheit) or lower to approximately 37 degrees.

For Further reading . . .

Only those books suitable for general reading have been listed, but those marked " S " are suitable for any who require the simplest possible approach.

Climate and Weather, by H. N. Dickson. Thornton Butterworth, Home University Library, Seventh Impression 1932, 2/6d.

The Drama of Weather, by Sir Napier Shaw. Cambridge University Press, Fourth Impression 1940, 10/6d.

The Air and Its Mysteries, by C. M. Botley. D. Appleton-Century, 1940, 3.00 dollars. " S ".

The Weather Eye, by C. R. Benstead. Robert Hale, 1940, 8/6d. " S ".

This Weather of Ours, by Arnold B. Tinn. George Allen & Unwin, 1946, 10/6d. " S ".

British Weather, by Stephen Bone. Collins, Britain in Pictures Series, 1946, 5/-. (Pictorial) " S ".

Climate Makes the Man, by Clarence A. Mills. Victor Gollancz, Third Impression March 1946, 7/6d. " S ".

The Book about Weather, by A. J. Mee. Littlebury & Co., The Worcester Press, 1947, 12/6d. " S ".

Weather Warnings for the Novice, by Percy Woodcock. Frederick Muller, Third Impression 1947, 4/-. " S ".

Here is the Weather Forecast, by E. G. Bilham. Golden Galley Press, Binnacle Books Series, 1947, 10/6d. " S ".

The Elements Rage, by Frank W. Lane. Country Life, Second Impression 1948, 10/6d. " S ".

Teach Yourself Meteorology, by " Aeolus." English Universities Press, Second Impression 1949, 4/6d.

Your Weather Service, prepared by the Central Office of Information for the Meteorological Office. H.M.S.O., 1950, 1/-. " S ".

Climate in Everyday Life, by C. E. P. Brooks. Ernest Benn, 1950, 21/-.

Weather Lore, by Richard Inwards. Rider and Co., Fourth Impression 1950, edited and revised by E. L. Hawke, 15/-. (Probably the world's largest collection of weather sayings, but many of them are inaccurate.) " S ".

The Weather, by George Kimble. Penguin Company, Pelican Series, Second (and enlarged) Edition 1951, 3/-. " S ".

Climate and the British Scene, by Gordon Manley. Collins, The New Naturalist Series, 1952, 25/-.

GAZETTE PRINTING SERVICE
BIDEFORD GAZETTE LIMITED

Enquiries sought for Printing of Handbooks of all kinds. Modern type faces and attractive results. Printers of this delightful booklet

5, GRENVILLE STREET, BIDEFORD
Telephone : Bideford 777

Visit

Bideford & Westward Ho!

Enchanting
holiday spots that
inspired Charles Kingsley
the famous novelist . . . For all
Books of local interest, address enquiries
to the publishers

Bideford Gazette Ltd.
" Bideford-in-Devon "
Telephone 777

ALLOW US TO INTRODUCE YOU TO A HERETIC!

ONE OF OUR AUTHORS IS A HERETIC! He is Captain C. E. Cookson, C.M.G., a former Acting Governor of Sierra Leone and a man who has enjoyed a long and adventurous career.

We reminded Captain Cookson recently that there was a time when heretics and their writings finished up at the stake. Or, alternatively, at the block. It may sound ridiculous now, but a man who told the Pope four centuries ago that the world was round had his head cut off!

That man "held an opinion contrary to authorised teaching," which is the way the dictionary defines a "heretic." But the man was right. Heretics often are right; just as often, in fact, as the authorised teachings are wrong.

What is the position regarding **Captain Cookson's** heresies? In his new book **A HERETIC'S ANSWER TO COMMUNISM** the author states that there is "something rotten in the State of Britain"—to quote the words of a reviewer. There is a great deal wrong with our present form of democracy, and a New and better Democracy is the only answer to Communism.

Captain Cookson disagrees with many of our policies, past and present, with regard to both home and Commonwealth affairs. He gives proposals which, although certainly not the "last word" on these matters, are intended to point the way to a beginning of better government.

A HERETIC'S ANSWER TO COMMUNISM is available through your bookseller or from The Western Press, Westward Ho! N. Devon, at 7/6d. It is a book which demands the attention of every thinking person.

The author is now just completing his Memoirs, which will be well illustrated. These are written in lighter vein and are entitled **MANY UPS AND DOWNS**—the many "ups and downs" of a very varied and adventurous career. This interesting book will be published at 15/-. Reserve your copy now! It will be sent to you immediately on publication.

LUNDY STAMPS

Complete your collection of these attractive Puffin Stamps from Lundy

1929-39 Puffins (8)..4/6
1951 Seabirds (7)..3/6
1953 Coronation (7) 4/-
1954 Air Pictorials (6)3/-

1954 Jubilee (island views) (7) 3/9
Postally used flown covers........1/6
P.C. Maps of Lundy 8d.
Hand-coloured Maps 8in. x 10in. 5/-

Extensive stocks of flown covers, varieties commemoratives, proofs and errors.

APPROVAL SELECTIONS ON REQUEST

Special Mounted Souvenir Selection 10/- post free

ATLANTIC COAST STUDIOS (Philatelic Department)
TOWER HOUSE, THE STRAND, BIDEFORD, DEVON

EARTHQUAKES, FIRE, WAR OFF THE AFRICAN COAST, STORMS GALORE, AND A SPECTACULAR LIFE IN THE "GOOD OLD DAYS" OF SAIL . . .

HERE IS THE TRUE STORY OF A SEAMAN'S LIFE!

in

Channels, Cloves and Coconuts

By Commander C. J. CHARLEWOOD
O.B.E., D.S.C., R.D., R.N.R. (Retd.)

Printed in Great Britain
by Amazon